Claire E. Flynn

The Rosen Publishing Group's
PowerKids Press™
New York

Published in 2009 by The Rosen Publishing Group, Inc.
29 East 21st Street, New York, NY 10010

Book Design: Haley W. Harasymiw

Photo Credits: Cover, p. 6 © ArchMan/Shutterstock; p. 5 (washer) © mehmetsait/Shutterstock; p. 5 (folded towels) © Joseph/Shutterstock; p. 5 (towel on rack) © Konovalikov Andrey/Shutterstock; p. 5 (dirty towels) © Feng Yu/Shutterstock; p. 7 © Vladimir Vitek/Shutterstock; p. 9 (water fountain) © Elena Elisseeva/Shutterstock; p. 9 (windsurfers) © Kochneva Tetyana/Shutterstock; p. 10 © Markus Botzek/Zefa/Corbis; p. 12 © Iakov Kalinin/Shutterstock; p. 13 © Valery Potapova/Shutterstock; p. 14 © Yellowj/Shutterstock; p. 15 © POMAH/Shutterstock; p. 16 © Stephen Bonk/Shutterstock; p. 17 © Marilyn Barbone/Shutterstock; pp. 18–19 (notebook) © David Good/Shutterstock; pp. 18–19 (cups of water) by Michael Flynn; p. 20 © Bogdan Shahanski/Shutterstock; p. 21 © Peter Wey/Shutterstock; p. 22 © Alan Heartfield/Shutterstock; p. 24 © Alex Staroseltsev/Shutterstock; p. 25 © LianeM/Shutterstock; p. 26 © Marekuliasz/Shutterstock; p. 27 © Sandra Zuerlein/Shutterstock; p. 29 © David W. Hughes/Shutterstock.

Library of Congress Cataloging-in-Publication Data

Flynn, Claire E.
Water world : earth's water cycle / Claire E. Flynn.
p. cm.
Includes index.
ISBN 978-1-4358-0193-6 (pbk.)
6-pack ISBN 978-1-4358-0194-3
ISBN 978-1-4358-2999-2 (library binding)
1. Hydrologic cycle—Juvenile literature. I. Title.
GB848.F59 2009
551.48-dc22

2008046766

Manufactured in the United States of America

CONTENTS

WHAT IS A CYCLE?

A cycle is a series of events that happens in the same order over and over. There are many different types of cycles in our everyday lives. To better understand the idea of a cycle, you may want to consider your clothes and how they fit into the laundry cycle.

Around and Around

First, you wear clean clothes. Then you put dirty clothes into a basket where they wait to be washed. Next, the clothes are washed and dried. After that, you fold the clothes and put them into a drawer, and the cycle begins again. The clothes move from your drawer, to your body, to the dirty-clothes basket, to the washer, to the dryer, and back to your drawer, where they're ready to be worn again. Next time you put on a clean shirt, you can think about the cycle it went through. This is just one of many cycles that take place in our lives.

Earth's Cycles

Earth has many cycles that help it stay healthy, **recycle** materials, and support life. The rock cycle, the nitrogen cycle, and the carbon cycle are three examples. Just like the laundry cycle, Earth's cycles include many events that occur over and over.

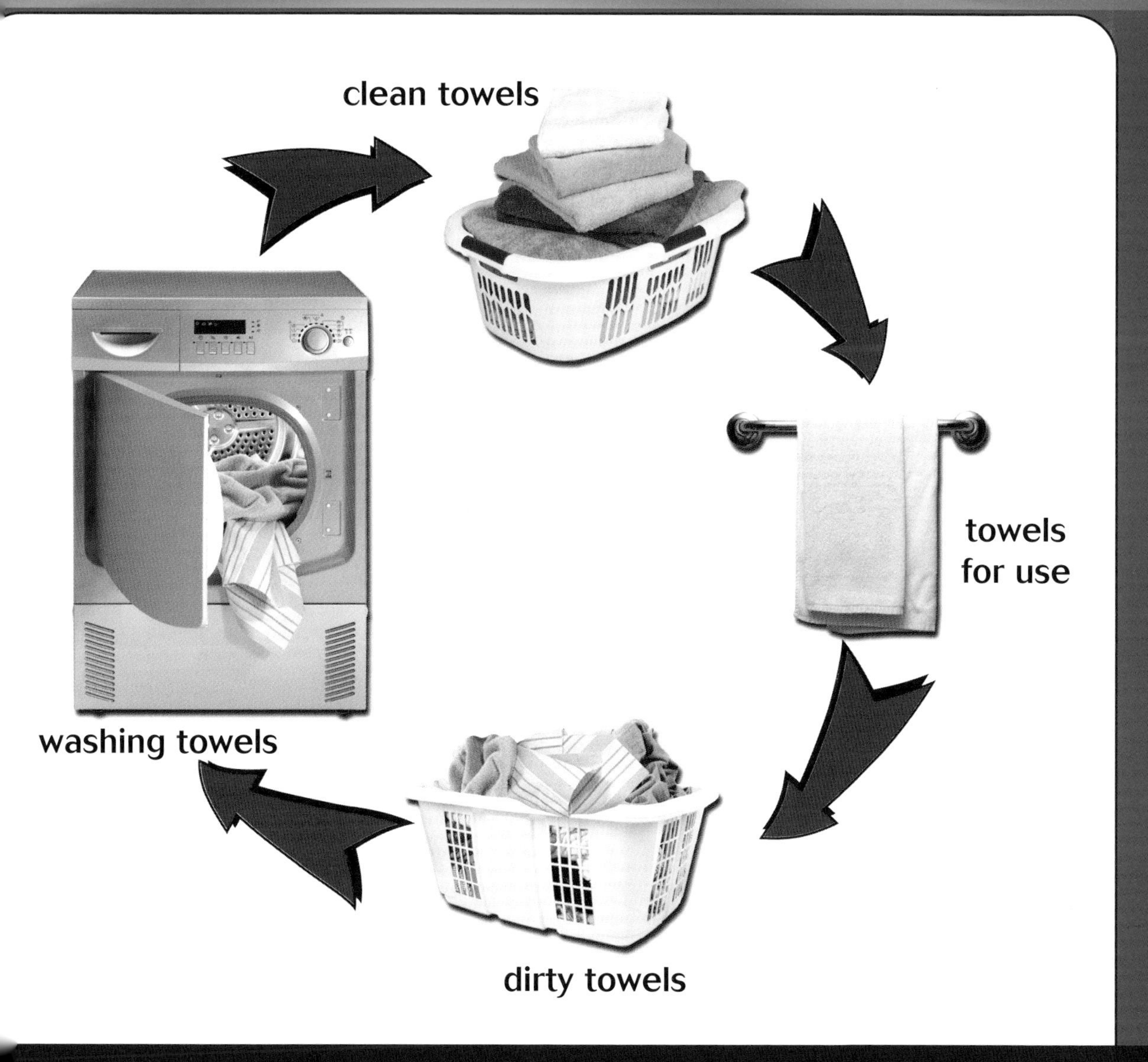

There are so many different types of cycles. Can you think of any other examples in your own life? You may be surprised to discover how many cycles occur in your home and at school.

In the rock cycle, rocks are recycled into other rocks! There are three types of rocks: sedimentary, igneous, and metamorphic. Rocks are constantly fusing together, breaking apart into smaller pieces, and forming new rocks.

All creatures need nitrogen to live. As a gas, nitrogen is abundant, but most plants and animals can't use it in this form. During the nitrogen cycle, lightning and special **bacteria** convert the gas into compounds in the soil for plants to use. Animals get nitrogen by eating

plants. When plants and animals die and decompose, or break down, they return nitrogen to the atmosphere.

The carbon cycle involves the atmosphere and Earth's interior. Plants remove carbon dioxide from the atmosphere during **photosynthesis**. The carbon is transferred into animals that eat the plants. When the animals and plants die, some carbon is released back into the atmosphere. However, some carbon dioxide from dead plants can be locked beneath the ground for millions of years. This is how fossil fuels like coal and oil form.

During the carbon cycle, the carbon dioxide in fossil fuels is released back into the atmosphere when we burn them. Too much carbon dioxide in our atmosphere is bad for Earth.

THE WATER CYCLE

Earth's water is constantly in motion. The water that was on the planet long ago is the same water on the planet today. The water you drink from a drinking fountain could be the same water that splashed onto the deck of Christopher Columbus's ship over 500 years ago! Actually, Earth has been recycling its water for more than 3 billion years. This process of recycling water is known as the water, or hydrologic, cycle.

How Does It Work?

Like all cycles, Earth's water cycle has no beginning or ending point. The sun, however, is a key part of the water cycle, so it's a good place to start to understand this process. The sun heats up Earth's oceans, lakes, and other bodies of water. Rising air currents take water vapor—water in the form of a gas—into

What Does It Mean?

The word "hydrologic" is the adjective form of "hydrology," which is a combination of two Latin words that were borrowed from the Greek language. *Hydro* means "water," and *logy* means "the study of." So hydrology is the study of Earth's water.

the atmosphere. This is called **evaporation**. Cool temperatures high in the atmosphere cause water vapor to change back into a liquid. This is called **condensation**. Soon clouds begin to form. As air currents blow clouds around the atmosphere, tiny drops come together to form larger drops.

On average, there are about 34 **trillion** gallons (128 trillion l) of water above your head each day! That's a lot of water! However, it's only a tiny part of Earth's total amount of water.

We need Earth's water to drink and water crops, but we also use it for transportaion and sport.

Precipitation falls from the clouds in the form of rain, snow, sleet, or hail. Much of the precipitation returns directly into oceans and lakes. The precipitation that falls on land seeps into the soil and between rocks. This water either evaporates once again, is used by plants, or becomes runoff, which is water that flows downhill into rivers, lakes, and oceans.

Like the example of the laundry cycle, the water cycle repeats itself. The water evaporates from the oceans, lakes, and other bodies of water, and then the water vapor condenses and turns into clouds. After that, precipitation falls from

During a rainstorm, runoff can turn trickling creeks into swift rivers.

the clouds. Precipitation and runoff refill the oceans and lakes. The cycle happens over and over again!

Water We Can Use

Living things can't survive without water. Almost three-quarters of Earth is covered with water, yet less than 1 percent of this water is available for people, plants, and animals to use. The oceans are filled with salt water. Most freshwater is locked up in ice caps and glaciers. This is why Earth's water cycle is so important. Let's take a closer look at the parts that make up the water cycle.

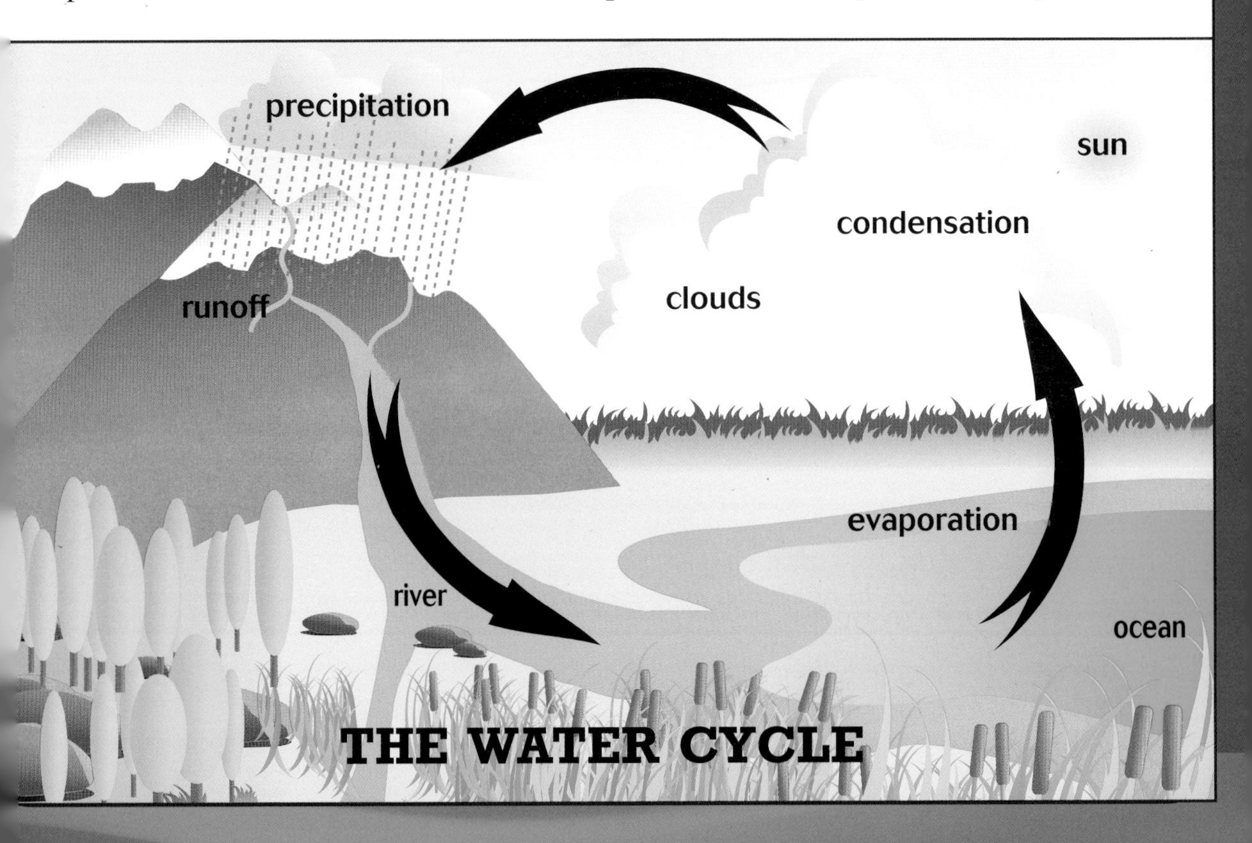

THE SUN

The sun is a very important part of the water cycle. The heat and energy from the sun warm the oceans and other bodies of water, causing evaporation. The sun's heat also causes water to evaporate from plants and animals. The evaporated water leads to water vapor in the air that rises and condenses to form clouds.

Did you know that the sun is partially responsible for the weather, too? The sun's heat and energy warm the oceans in some parts of the world. Due to the lack of direct sunlight in other parts of the world, the oceans there are cooler. This cooling and warming of the water causes ocean currents that move massive amounts of the cold and warm water around.

Not only are ocean currents caused by the sun, but they are also partially responsible for air currents in the atmosphere. The sun heats up the land and the air above it, which causes air to rise and fall. It also creates wind, which can move clouds from one area to another. Next time you're sitting on the beach enjoying the sun and the warm breeze, you'll know they go hand in hand.

Along with the moon, the sun also plays a role in the cycle of tides in Earth's oceans.

EVAPORATION AND CONDENSATION

Evaporation occurs when heat changes water from a liquid to a gas. Have you ever seen a puddle in the morning, only to find that it had disappeared after being in the sun all day? This is because of evaporation. Just like water in a puddle, water in lakes, rivers, and oceans evaporates into the atmosphere. Evaporation is the main source of water in the atmosphere.

Water in gas form is called water vapor. Steam from a cup of hot water is water vapor that you can see. Next time you're outside on a cold winter day, see what happens when you breathe into the air. If it's cold enough, you'll see water vapor coming from your mouth and nose. The air we breathe out has water vapor in it. When you hear

The steam that's produced by hot water is also very ho[t]. Don't touch it or you'll get burned.

FUN FACT

Sometimes the snow and ice in Earth's ice caps and glaciers turn back into water vapor without melting first. This is called sublimation.

someone say it's humid, that means the air has a lot of water vapor in it. It's usually more humid on hot days. This is because heat causes water to evaporate and turn into water vapor. Warm air can also hold more moisture than cold air.

Do Plants Sweat?

Water can also evaporate directly from plant leaves and stems. The process of plants giving off water directly into the atmosphere is called

transpiration. To help you remember the word *transpiration*, think of human **perspiration**. Just like you, plants sweat to cool off. Similar to sweat evaporating off your skin, water evaporates from the leaves and stems of plants during transpiration.

A very big oak tree can transpire 40,000 gallons (151,400 l) of water per year! Transpiration is responsible for about 10 percent of the water vapor in the air.

Condensation

Condensation is the opposite of evaporation. Condensation is the process of water vapor in the air changing from a gas into a liquid when it gets cooler. Water vapor condenses to form clouds and fog. It can also result in little water droplets forming on things, such as the dew we see on plants on a cool morning.

Have you ever heard someone say that a glass of water is sweating when it has little water droplets on its outside surface? The water doesn't seep through the glass; the droplets result from condensation. The warm air around the glass

The mist and rain clouds in this picture were both caused by condensation.

cools down because of the cold liquid inside the glass. This changes the water vapor in the air into water droplets on the outside of the glass.

You will need:
- sunny window
- clear plastic cup
- water
- marker
- pencil and paper

See for Yourself

This experiment will help you see evaporation at work. Fill a clear plastic cup with water as high as you'd like. Next, use a marker to draw a line on the outside of the cup at the water level. Leave the cup by the window and watch what happens.

As time goes on, you'll notice that the water level drops due to evaporation. The sunlight coming through the window heats

the water in the cup. The water slowly changes from a liquid into a gas. The water vapor then becomes part of the air you breathe. Make a chart to keep track of how far the water level goes down daily.

Do you want to see condensation at work? Take the plastic cup and fill it with ice water. Let it stand in a sunny window for a while. What happens to the outside of the cup? Before too long, you'll notice tiny water droplets starting to form on the outside of the cup. This is because the warm air around the cup is cooled by the ice water inside the cup. Water vapor in the air becomes water droplets on the cup. This shows water vapor turning back into a liquid.

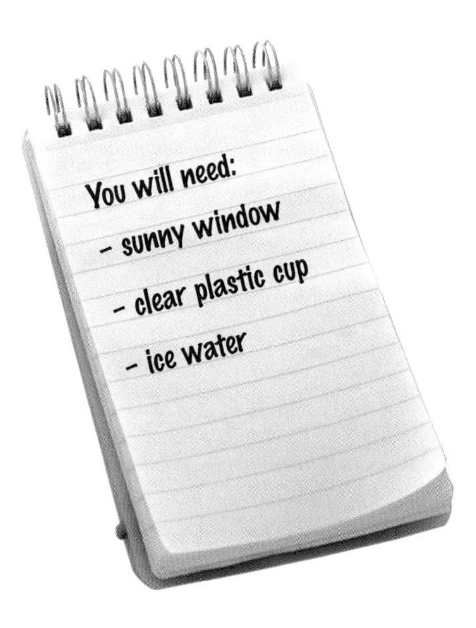
You will need:
- sunny window
- clear plastic cup
- ice water

CLOUDS

The air is full of water. A lot of the time, we can't see the water vapor in the air because the particles are just too small to be seen. The warm air that's close to the ground contains water vapor. As this warm air rises, it cools. Cool air doesn't hold as much water as warm air, so the water vapor condenses to form water droplets. Condensation is what forms clouds. So, a cloud is actually water in the atmosphere that we can see!

The water droplets cling to tiny dust particles in the atmosphere. If it's cold enough, some of the water droplets turn into ice crystals. Both the ice crystals and the water droplets float in the air and cling together, forming the clouds we see in the sky.

Have you ever noticed that some clouds are darker than others? This is because white clouds have tiny droplets packed close together. Most light that hits the clouds is reflected, which makes them look white. Dark clouds have larger drops that are spaced farther apart. Less light reflects off the clouds, making them appear darker.

Clouds are a very important part of the water cycle because they provide the precipitation that refills Earth's rivers, lakes, and oceans.

PRECIPITATION AND RUNOFF

When the water droplets that form clouds become too heavy to stay in the air, they fall as precipitation. Most precipitation falls as rain. Some of the precipitation that falls as snow lands on glaciers or ice caps and remains frozen for many years.

Runoff

Besides precipitation, runoff is another way water finds its way back to rivers, lakes, and oceans. After a heavy rain, you can sometimes see runoff when

Snowmelt often fills rivers and lakes quickly, sometimes causing spring floods.

it streams down your driveway or down a hill. If you live in an area with cold winters, you are probably familiar with the runoff caused by melting snow. Snowmelt, which is a major part of the water cycle, occurs in some areas at the end of winter. As winter snow melts, it turns into water that becomes runoff.

What Is an Aquifer?

Not all the water that falls on land becomes runoff. Much of it soaks into the ground like water into a sponge. Water that seeps deep into the ground helps refill **aquifers**. Aquifers store large amounts of freshwater for a long time. The top surface of an aquifer is called a water table. Wells are often drilled into aquifers, making it possible for water to be pumped out for people to use. For many people, aquifers are the only source of freshwater.

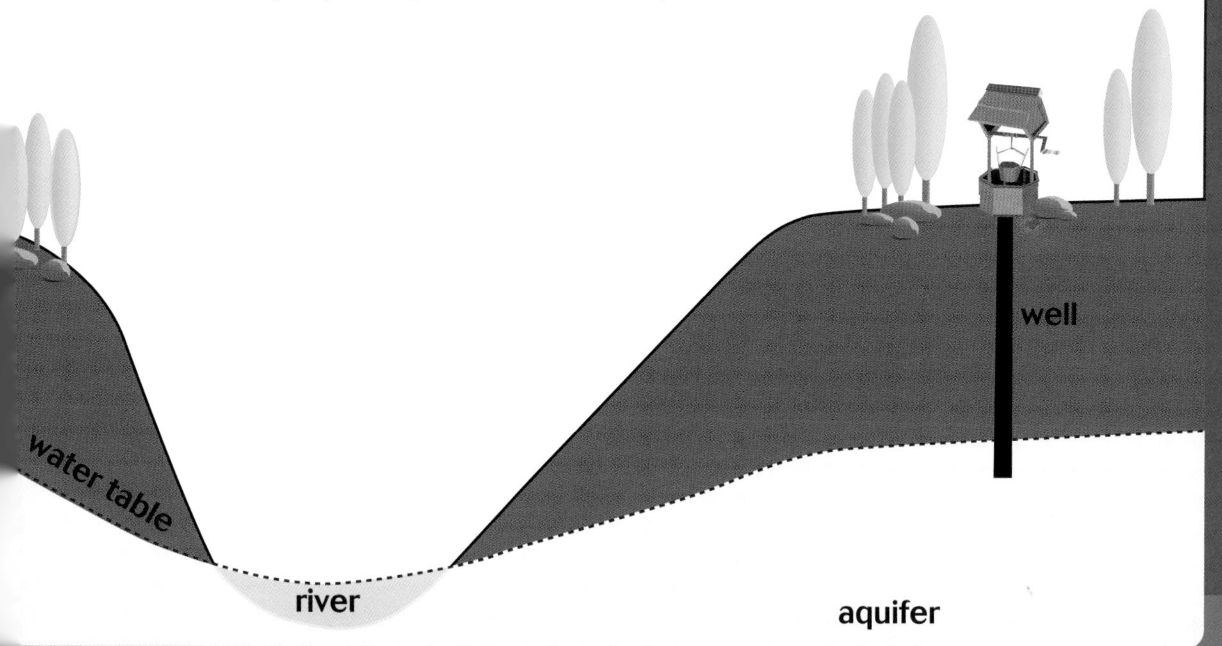

OCEANS

Most of Earth's water is stored in the oceans for long periods of time. It's often difficult for us to grasp how vast the oceans are. The average depth of the oceans is 12,200 feet (3,700 m). However, the deepest part of the oceans is almost 7 miles (11 km) below sea level. To give you an idea how deep that is, Earth's tallest mountain—Mount Everest—is only about 5.5 miles (9 km) tall! During a 100-year period, the amount of water in Earth's oceans doesn't change very much. Throughout Earth's history, however, climate changes have greatly affected the oceans' water levels. Thousands of years ago, sea level was lower because Earth's climate was colder. Much of Earth's water was frozen in glaciers and ice caps. During the last **ice age**, Earth had so many glaciers that sea level was 400 feet (122 m) lower than it is today. The opposite is true during very warm periods. Millions of years ago, when it was much warmer on Earth, sea level may have been 655 feet (200 m) higher than it is today.

About 97 percent of the water on Earth is salt water. The amount of salt in water is referred to as salinity.

FRESHWATER

Lakes, rivers, and **wetlands** are major sources of freshwater that are available to people. Unlike oceans, freshwater doesn't contain salt. Freshwater may be stored in the ground. Groundwater is the largest source of unfrozen freshwater. Some people dig wells that provide them with all the water they need for their homes. Much freshwater is stored in the form of ice in glaciers. However, this water is not easily used because it is frozen.

Plants and animals depend on freshwater to live. People need to drink water to replace the water we lose during the day. Without freshwater, our bodies would not be able to function properly. Farmers use a great deal of

This is a freshwater lake in Colorado. When the gates are opened, an irrigation canal takes water to farms, where it is used to water crops.

freshwater to water crops that we eat. Many factories also depend on freshwater to make their products. These are just a few uses for freshwater. Think about all the ways you use water in one day!

As you read in Chapter Two, more than 70 percent of Earth is covered by water. You would think there would be plenty to go around. However, the majority of Earth's water is salt water, so we must take care of the freshwater we have!

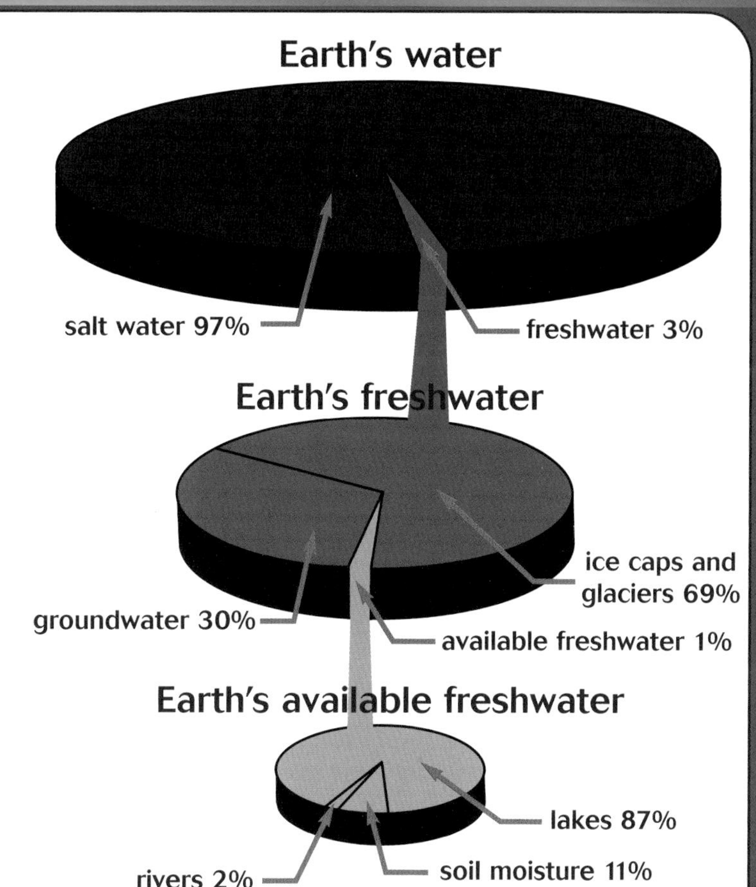

HOW YOU CAN HELP THE WATER CYCLE

Although the water on Earth is continually being recycled, the amount of freshwater available to us is limited. There are many things we can do to keep our water supply safe and clean.

Conserving Water

Take a look at your daily activities. Can you think of ways to decrease the amount of water you use? One easy way is to cut down the time you use to shower. Another simple way is to not leave the water running while you're doing dishes or brushing your teeth. Run the dishwasher only when it's full. Another way to save water in your home is by fixing leaky faucets and pipes. Those small drips add up!

Do you like to garden? Instead of using a hose to water flowers, try collecting rainwater in buckets on the sides of your house. It's a free and easy way to water your plants. Another way to conserve water is by sweeping driveways and sidewalks rather than using a hose. Next time your parents wash the car, you may want to suggest taking it to the car wash, which uses roughly 45 gallons (170 l) of water per wash. If you wash your car with a hose at home, you may use up to 140 gallons (530 l) of water!

A dripping faucet can waste 20 gallons (76 l) of water per day. A leaky toilet can waste 200 gallons (757 l) per day!

Clean Water

In addition to conserving water, we should also be careful not to **contaminate** our freshwater supply. Always be aware of what is being put down your drains at home. Toxic chemicals, such as those found in some soaps and cleansers, can contaminate our freshwater supply. Just 1 gallon (3.8 l) of oil can contaminate 1 million gallons (3,785,000 l) of freshwater! There are proper ways to dispose of things such as used oil. Take the steps to find out how.

We can work together to conserve and protect Earth's water. By following the advice in this chapter, we can help keep our rivers, lakes, and aquifers clean. Can you think of some other ways we can help keep Earth's water cycle healthy?

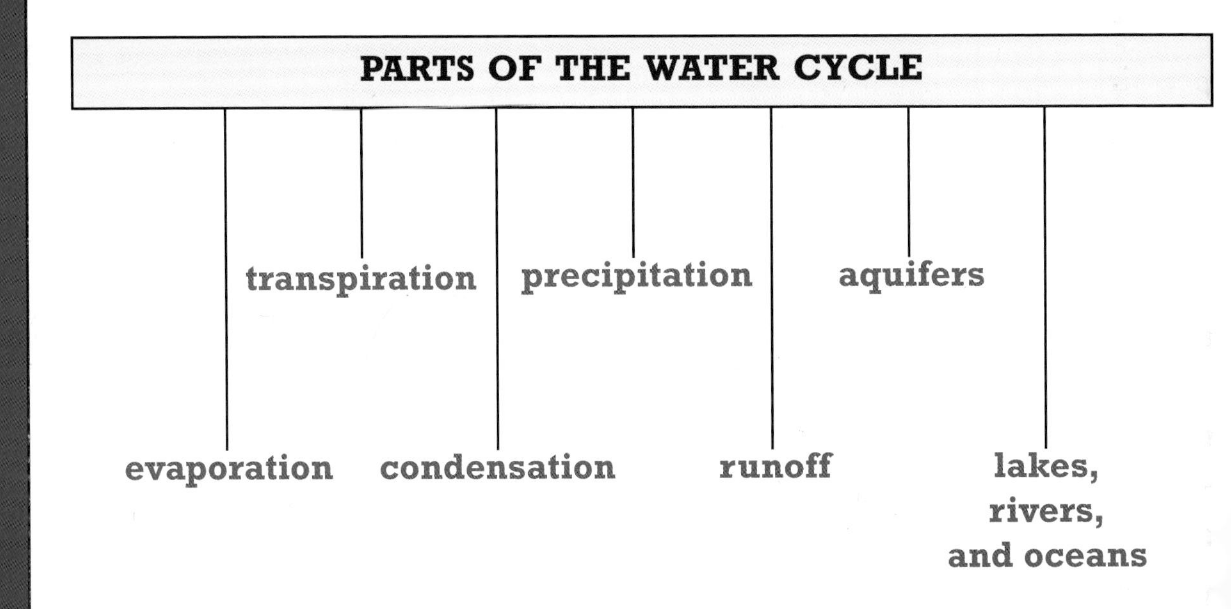

GLOSSARY

aquifer (AH-kwuh-fuhr) A layer of rock, sand, or gravel underground that holds water.

bacteria (bak-TIHR-ee-uh) Tiny living things that can't be seen with the eye alone. Some cause illness, but others are helpful.

condensation (kahn-dehn-SAY-shun) The process that changes a cooled gas into drops of liquid.

contaminate (kuhn-TA-muh-nayt) To make something unusable by adding toxins to it.

evaporation (ih-vaa-puh-RAY-shun) The process that changes a liquid into a gas.

ice age (EYS AYJ) A period of time when ice and glaciers covered large parts of the land.

perspiration (puhr-spuh-RAY-shun) Sweat.

photosynthesis (foh-toh-SIHN-thuh-suhs) The way in which green plants make their own food from sunlight, water, and carbon dioxide.

precipitation (prih-sih-puh-TAY-shun) Any moisture that falls from the sky, such as rain and snow.

recycle (ree-SY-kuhl) To pass something through a series of steps, such as water in the water cycle.

transpiration (trans-puh-RAY-shun) Evaporation from the leaves and stems of plants.

trillion (TRIHL-yuhn) One thousand billions (1,000,000,000,000).

wetland (WEHT-land) Land that is either covered with water most of the time or has a lot of moisture in it.

INDEX

Due to the changing nature of Internet links, The Rosen Publishing Group, Inc., has developed an online list of Web sites related to the subject of this book. This site is updated regularly. Please use this link to access the list: http://www.rcbmlinks.com/rlr/water